PACIFIC URBAN DEVELOPMENT, WATER, AND SANITATION SECTOR ROAD MAP 2021–2025

APRIL 2021

ASIAN DEVELOPMENT BANK

ADB

© 2021 Asian Development Bank
6 ADB Avenue, Mandaluyong City, 1550 Metro Manila, Philippines
Tel +63 2 8632 4444; Fax +63 2 8636 2444
www.adb.org

Some rights reserved. Published in 2021.

ISBN 978-92-9262-810-9 (print); 978-92-9262-811-6 (electronic); 978-92-9262-812-3 (ebook)
Publication Stock No. SGP210155-2
DOI: http://dx.doi.org/10.22617/SGP210155-2

The views expressed in this publication are those of the authors and do not necessarily reflect the views and policies of the Asian Development Bank (ADB) or its Board of Governors or the governments they represent.

ADB does not guarantee the accuracy of the data included in this publication and accepts no responsibility for any consequence of their use. The mention of specific companies or products of manufacturers does not imply that they are endorsed or recommended by ADB in preference to others of a similar nature that are not mentioned.

By making any designation of or reference to a particular territory or geographic area, or by using the term "country" in this document, ADB does not intend to make any judgments as to the legal or other status of any territory or area.

Please contact pubsmarketing@adb.org if you have questions or comments with respect to content, or if you wish to obtain copyright permission for your intended use that does not fall within these terms, or for permission to use the ADB logo.

Corrigenda to ADB publications may be found at http://www.adb.org/publications/corrigenda.

Notes:
In this publication, "$" refers to United States dollars.
ADB recognizes "Hong Kong" as Hong Kong, China.

On the cover: Stilt homes in Port Moresby, Papua New Guinea (photo by ADB).
All photos are from ADB.

Printed on recycled paper

CONTENTS

TABLES, FIGURES, AND BOXES

Tables

Figures

Boxes

FOREWORD

The vision underpinning this road map is for each intervention to build the resilience of people, institutions, and the environment through investments, knowledge, and partnerships. ADB's work in the Pacific urban sector supports its developing member countries in providing safe, efficient, and reliable urban services that drive equitable socioeconomic growth, build resilience to shocks, and achieve sustainable results. ADB's work in the Pacific urban sector focuses on three main pillars: (i) improved urban services (including water supply, sanitation, hygiene, solid waste management, and urban flood management); (ii) spatial planning; and (iii) climate change and the environment. The road map includes support along technical, financial, social, and governance dimensions to help achieve greater impact and more sustainable outcomes.

By supporting greater use of technology, knowledge solutions, institutional strengthening, private sector participation, and the building of social capital, ADB seeks to help countries in the Pacific region build strong, healthy, and inclusive economies, while mitigating the challenges associated with their geographic isolation, limited resources, and vulnerability to climate events.

Leah C. Gutierrez
Director General
Pacific Department
Asian Development Bank

ACKNOWLEDGMENTS

This road map, 2021–2025, presents the vision and targets of ADB for the urban development, water, sanitation, and related subsectors in the Pacific. It aims to guide ADB project and program design and implementation. It has been developed based on the ADB team's experiences, lessons, and aspirations, and through ongoing dialogue with Pacific developing member countries (DMCs) on their challenges and priorities.

The document is intended to support ADB in communicating its vision for the urban development, water, sanitation, and related subsectors with Pacific DMCs and development partners.

The road map was prepared by Vivian Castro-Wooldridge, senior urban development specialist, ADB Pacific Department with inputs from Jingmin Huang, director, Urban Development, Water Supply and Sanitation Division, ADB Pacific Department. This road map benefited from the inputs and reviews of the ADB Pacific Department team including Stephen Blaik, principal urban development specialist; Kristina Katich, senior urban development specialist; Anupma Jain, senior urban development specialist; Yuki Ikeda, public management economist; and Alexandra Conroy, urban development specialist. It has further benefited from inputs provided by Noelle O'brien, principal climate change specialist, and Mairi Macrae, social development specialist (gender and development), of ADB's Pacific Department; and Lara Arjan, urban development specialist, and Sunghoon Kris Moon, urban development specialist, of ADB's Sustainable Development and Climate Change Department.

The production of the road map was supported by Cecilia Caparas, associate knowledge management officer; Raymond De Vera, senior operations assistant; and Herzl Banacia, operations assistant, of ADB's Pacific Department. This guide was edited by Ricardo Chan and Lawrence Casiraya and the layout and design was executed by Alfredo De Jesus.

Stilt homes in Port Moresby, Papua New Guinea.

The Asian Development Bank has ongoing or recently completed projects in all of its 14 Pacific developing member countries (DMCs), except for the Cook Islands, Nauru, Niue, and Samoa.[1]

Fragility. The cost of providing services in these countries is high because of the small size and isolation of most Pacific DMCs. Achieving sustainable and inclusive growth is challenging as the countries are extremely fragile and vulnerable to climate change impacts and natural disasters, and have narrow economic bases.[2] Frequent natural disasters can be devastating, and have a disproportionate impact on women, the poor, and the vulnerable.[3] Globally, Pacific DMCs account for 4 of the 10 countries at highest risk of natural disasters.[4]

Each intervention shall build the resilience of people, institutions, the environment, and infrastructure through investments, knowledge, and partnerships.

Urbanization. The population living in urban areas ranges from 100% in Nauru, a single island of 21 square kilometers (km^2); to 13% in Papua New Guinea (PNG), with 600+ islands totaling 462,840 km^2.[5] Urbanization is on the rise in most countries because of rural–urban migration and high birth rates; and some of the capital cities have population densities comparable to Hong Kong, China or Tokyo.[6] Urbanization has resulted in many social, economic, and environmental issues, including the rise of noncommunicable diseases, environmental degradation, and an increase in unplanned settlements.

---

[1] Excludes regional technical assistance, which may cover all 14 Pacific DMCs. The 14 Pacific DMCs are the Cook Islands, the Federated States of Micronesia, Fiji, Kiribati, the Marshall Islands, Nauru, Niue, Palau, Papua New Guinea (PNG), Samoa, Solomon Islands, Tonga, Tuvalu, and Vanuatu.

[2] Excerpt from ADB. 2020. *Annual Pacific Urban Update*. Manila.

[3] ADB. 2012. *The State of Pacific Towns and Cities: Urbanization in ADB's Pacific Development Member Countries*. Manila.

[4] Bündnis Entwicklung Hilft and Ruhr University Bochum—Institute for International Law of Peace and Armed Conflict. 2019. *WorldRiskReport 2019*. Berlin. ADB categorizes the following seven Pacific DMCs as fragile and conflict-affected situation countries: Federated States of Micronesia, Kiribati, Marshall Islands, Nauru, PNG, Solomon Islands, and Tuvalu.

[5] The demographic and administrative data for each country, such as population, area of territory, and number of islands, are in Appendix 1.

[6] South Tarawa in Kiribati had a population density of 3,184 persons/km^2 in 2010. The island of Ebeye in the Marshall Islands has a population density of 40,000 person/km^2.

Urban water supply, sanitation, and hygiene. Pacific DMCs lag behind other regions in accelerating access to basic water supply and sanitation (WSS) with only a 4% increase in water supply coverage between 2000 and 2017 compared to 8% globally; and 3% compared to 17% globally for access to basic sanitation coverage.[7] While access to water supply and basic sanitation is relatively high in urban areas in Pacific DMCs, coverage is not keeping pace with urbanization. As a result, service coverage is stagnating or even decreasing as the urban population grows. High access figures do not necessarily translate into a high quality of services, e.g., in terms of water quality and reliability, and can mask inequities in access. For sanitation, there are few examples in the region of safely managed sanitation across the whole service chain (Figure 1; see country data in Appendix 1).

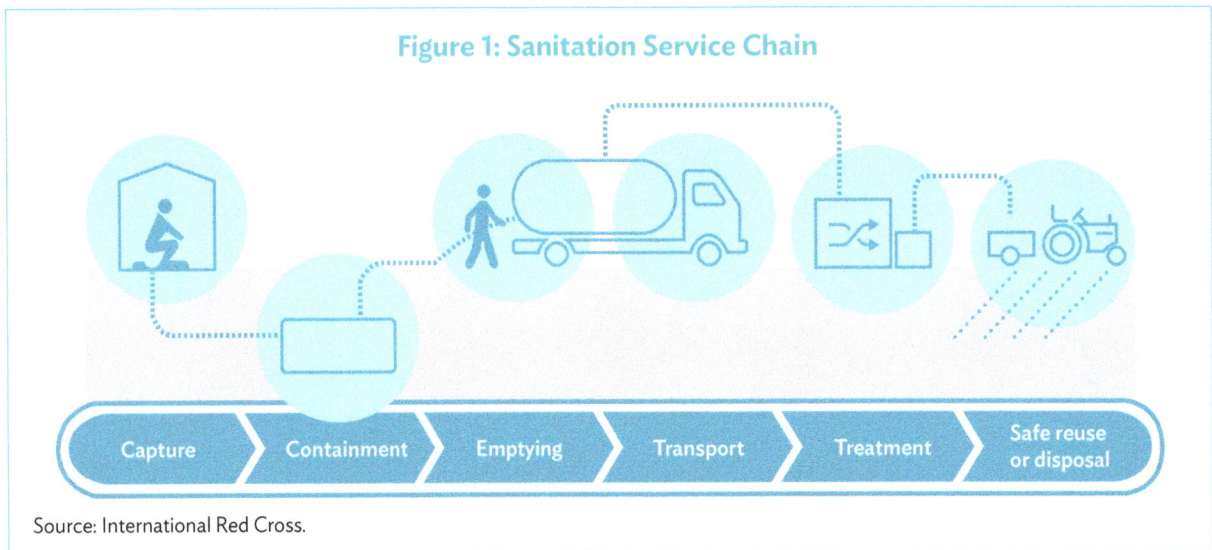

Figure 1: Sanitation Service Chain

| Capture | Containment | Emptying | Transport | Treatment | Safe reuse or disposal |

Source: International Red Cross.

---

[7]   United Nations Children's Fund (UNICEF) and World Health Organization (WHO) (Joint Monitoring Programme). 2010. *Progress on Household Drinking Water, Sanitation and Hygiene (2000–2017)*. Geneva. Figures are provided for all small island developing states (Figure 3).

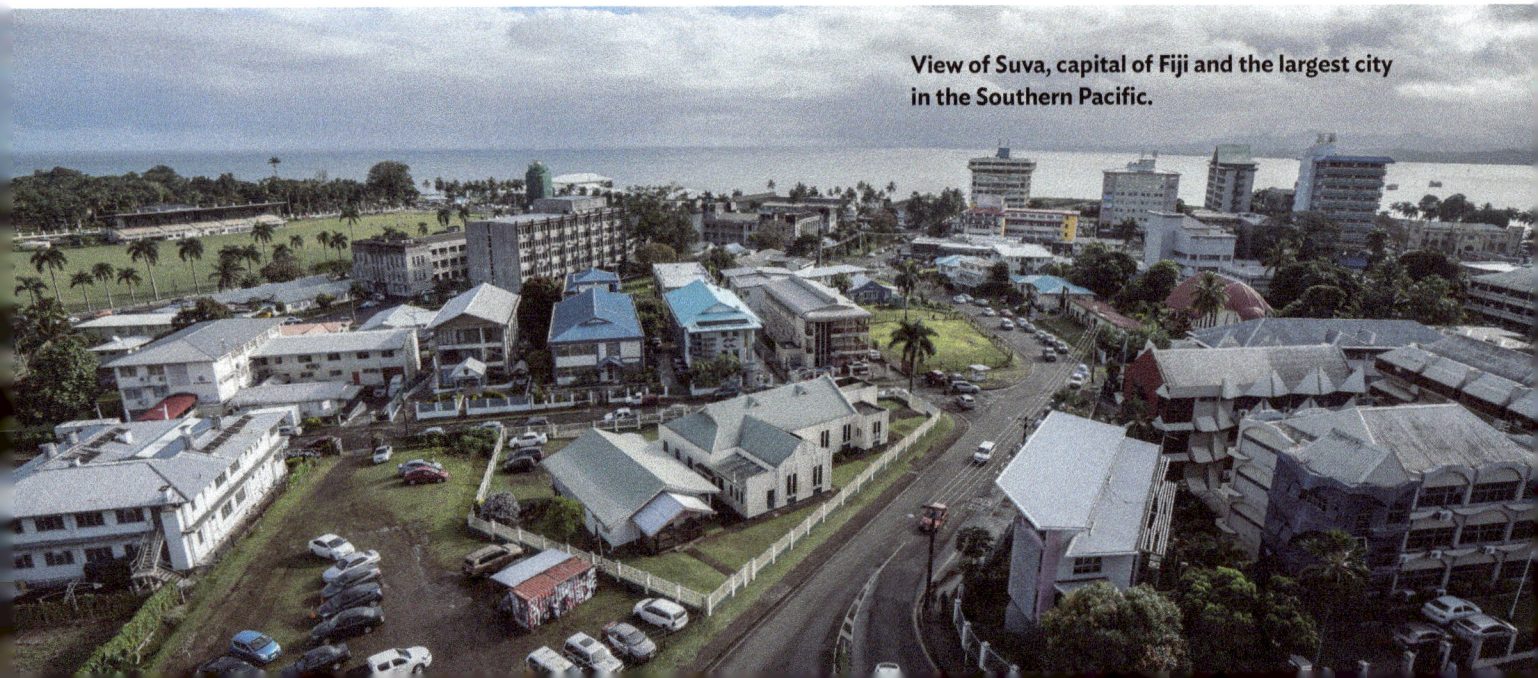

View of Suva, capital of Fiji and the largest city in the Southern Pacific.

Figure 2: Urban Sanitation Ladders by Region

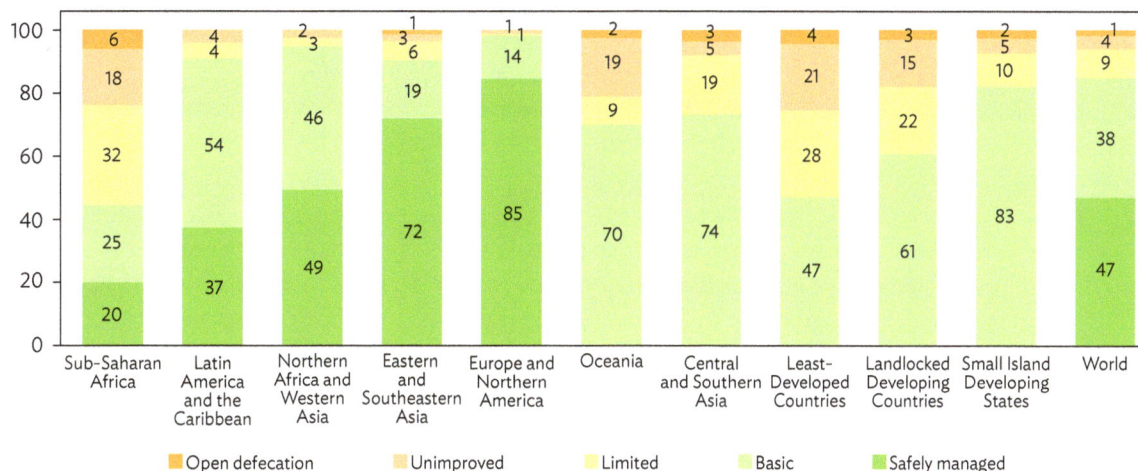

Source: Joint Monitoring Program 2019.

In addition to the water departments and local governments supplying WSS services, there are 29 water supply utilities and 20 utilities providing wastewater services.[8] Water supply coverage tends to be much higher than wastewater coverage, with only five utilities having the same coverage for both water supply and wastewater (Figure 3). Only 12 out of 20 utilities are treating the wastewater. Several Pacific DMCs, like Kiribati, the Marshall Islands, Nauru, Tokelau, and Tuvalu, have no reliable water resources. Cost of water production and energy

Figure 3: Wastewater Services of Pacific Water and Wastewater Association Members

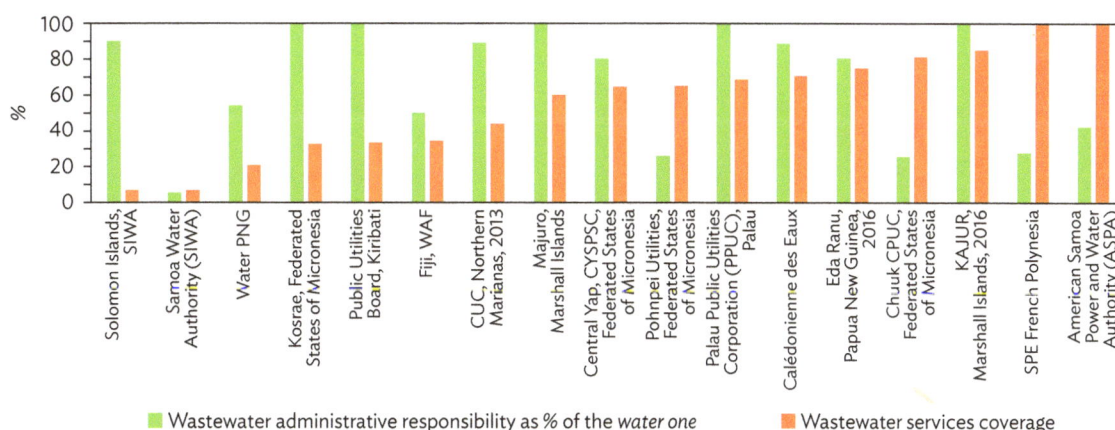

CUC = Commonwealth Utilities Corporation, CPUC = Chuuk Public Utilities Corporation, CYSPSC = Central Yap State Public Service Corporation, KAJUR = Kwajalein Atoll Joint Utility Resources, PNG = Papua New Guinea, SPE = Société Polynésienne des Eaux, WAF = Water Authority of Fiji.
Source: Pacific Water and Wastewater Association 2017 Benchmarking Report.

---

[8]   Pacific Water and Wastewater Association. 2018. *Benchmarking 2017 – Water Sector in Transition: 7 Years of Benchmarking*. Apia. The utilities also include non-ADB DMCs such as American Samoa and New Caledonia.

required is high, especially for those countries relying on desalination. Many islands rely on diesel generators for electricity generation, contributing to high operating costs. Many of the systems urgently need rehabilitation and modernization. Nonrevenue water levels exceed 40% in at least 16 PMDC utilities. Water supply, sanitation, and hygiene are often fragmented across different institutions. Efforts on sector coordination and monitoring of public health related to water supply, sanitation, and hygiene are not often sustained.

Solid waste management. Effective solid waste management (SWM) is a key intervention under ADB's Healthy Ocean Initiative.[9] It has direct benefits on the environment and also on health and economic development, such as for the tourism sector. Strategic areas for ADB support are holistic and integrated SWM covering the "4 Rs" (refuse, reduce, reuse, recycle), and entailing sustainable financing models, medical waste management, policy support, institutional strengthening, and raising awareness through behavior change campaigns. Significant support is still required to manage the range of waste types: domestic; commercial; institutional; industrial; medical; electronic; and other difficult-to-handle waste like scrap metal, disaster debris, and asbestos.

Urban development. Urban areas in the Pacific DMCs are atoll cities, coastal cities, or inland cities, with 90% of the population outside PNG, for example, living within 5 kilometers of the ocean. Therefore, the ocean is central to planning for more resilient and sustainable settlements. The ocean has social, cultural, and economic values, and is also essential for maintaining ecosystem services supported by the ocean. In some countries, the lack of affordable housing has led to the expansion of informal coastal settlements in areas that are prone to flooding and lack access to urban services.

---

9   Asian Development Bank. *Action Plan for Healthy Oceans.* https://www.adb.org/sites/default/files/am-content/484066/action-plan-flyer -20190430.pdf.

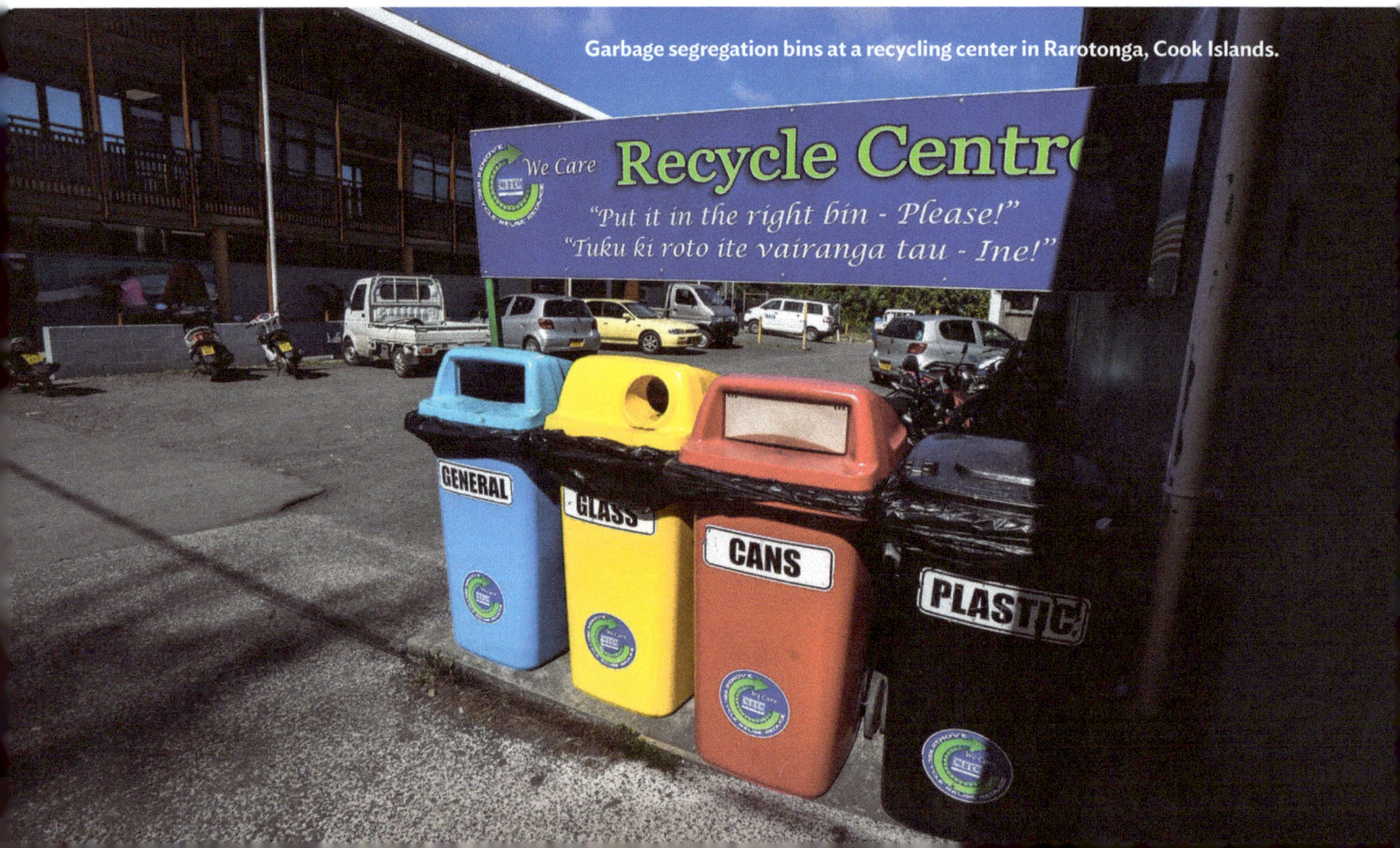

Garbage segregation bins at a recycling center in Rarotonga, Cook Islands.

2 MAJOR CHALLENGES

A dumpsite in Funafuti, Tuvalu.

The major challenges that need to be addressed are driven largely by the **interconnected climatic, social, and economic changes** that are transforming the natural and built environments at an unprecedented rate. These include:

(i)   **Rapid and unplanned urbanization** resulting in the growth of dense informal settlements or "urban villages" with insecure land tenure, lacking basic services, and often on sites vulnerable to disasters.[10]

(ii) **Increase in climate change impacts** due to greater frequency and intensity of severe climatic events resulting in damaging floods and droughts; and slow onset events like sea level rise affecting coastal lands and contaminating freshwater lenses. The vulnerable are most affected as they lack the resources to cope with water stresses or to build back; and Pacific DMCs' development funds are often diverted for emergency response, recovery, and rebuilding following a disaster.

(iii) **Scarcity of freshwater resources** because of the loss of watershed habitat, pollution from encroachment, and saline intrusion, even as the demand for freshwater increases.

(iv) **Limited access to basic sanitation and hygiene**, particularly in informal settlements with an impact on residents' health, especially children. Household septage tanks are often poorly constructed and the high density of informal settlements results in an unhygienic living environment. Poor sanitation also pollutes the environment, including groundwater that may be used for water supply. Poor menstrual health and hygiene are also prevalent and have a negative impact on girls' education.

(v)   **Increasing volumes of solid waste** as people move to cities and depend more on imported and processed foods and products and systems, and existing SWM systems are inadequate.

---

[10]   ADB. 2016. *The Emergence of Pacific Urban Villages: Urbanization Trends in the Pacific Islands*. Manila.

Pacific DMCs' capacity to build resilience and provide sustainable urban services for all, including the poor are constrained by the following:

(i)   **Fragility of the economies** as most Pacific DMCs do not have diversified industries or a deep economic base and are highly vulnerable to external shocks such as pandemics and rising commodity prices. Many Pacific DMCs depend significantly on tourism.

(ii)   **Geographic isolation** resulting in costly transportation of goods and materials. Many ships importing goods to Pacific DMCs return empty because of limited exports. Pacific DMCs also rely substantially on international consultants, contractors, and even skilled labor.

(iii)   **Limited public land for development** and customary land are sometimes untitled. There may also be conflicts over landownership or strong social and cultural ties to the land, making land acquisition for development difficult. Understanding of land-related challenges is important when considering ambitious development works or preparing urban development plans that may require land acquisition and/or resettlement. Significant consultations with the local population are required for any intervention with land implications in the Pacific DMCs, and these interventions are likely to take time.

(iv)   **Weak institutional capacity** affects the Pacific DMCs' ability to be forward-looking in terms of strategic directions and investment planning, and to implement multiyear plans. Policies and regulations may exist, but capacity to enforce them varies for both national and local governments. Weak capacity also impacts on the preparation and management of contracts that deliver high-quality and cost-effective civil works.

(v)   **Limited human and technical capacity** to carry out development work as the populations are relatively small and many professionals seek economic opportunities abroad.

(vi)   **Limited or no preventive maintenance programs for infrastructure** as many service providers lack the skills for asset management and maintenance and rely on subsidies for operation and maintenance (O&M), which may be inadequate or unreliable to ensure that assets perform over the expected lifetime.

(vii)   **Gender inequality** with social and gender norms that can limit women and girls' access to services and resources, economic empowerment, and participation and leadership. Persistent gender gaps remain, and there are high levels of gender-based violence.

Investments on urban infrastructure are not typically reflected as a priority in most of the Pacific DMCs' budgets. This may be because of the countries' emphasis on energy and transport infrastructure, and pressing needs brought about by climate changes and disasters.

3 VISION AND 5-YEAR TARGETS
Following ADB's Strategy 2030

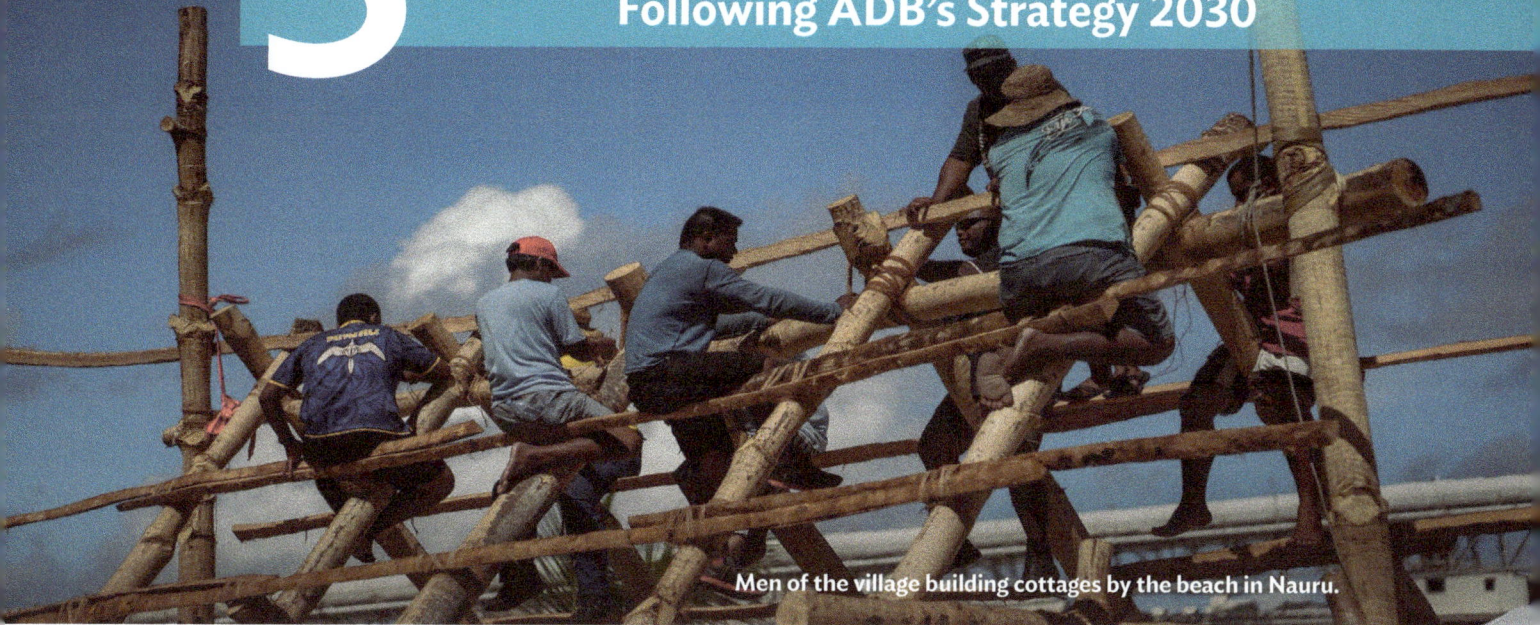

Men of the village building cottages by the beach in Nauru.

The vision and targets are designed to help achieve ADB's Strategy 2030, which promotes integrated solutions and a differentiated approach for small island developing states.[11] ADB's program in the Pacific will also be aligned with its overarching corporate objective to achieve a prosperous, inclusive, resilient, and sustainable Asia and the Pacific through (i) livable, inclusive, safe, and resilient cities; (ii) climate-resilient infrastructure; (iii) resource efficiency; and (iv) healthy oceans.[12] ADB shall strive to maintain its position as a key partner to Pacific DMCs for knowledge and investment on urban development through long-term partnerships; and will leverage the support of other development partners and donors for building resilience, introducing innovation, and scaling up successful interventions.

Box 1: The Differentiated Approach to Small Island Developing States in ADB's Strategy 2030

- Climate change adaptation, environmental sustainability, and disaster risk management

- Connectivity and access

- Institutional strengthening

- Efforts to improve business environment and promote private sector-led growth

Source: Asian Development Bank (ADB). 2018. *Strategy 2030: Achieving a Prosperous, Inclusive, Resilient, and Sustainable Asia and the Pacific.* Manila.

Our support will focus on improving improved urban services, improving spatial planning and focusing on climate change and the environment.

---

[11] Small island developing states present three key characteristics: (i) small size, (ii) remoteness and isolation, and (iii) a maritime environment.
[12] ADB. 2018. *Strategy 2030: Achieving a Prosperous, Inclusive, Resilient, and Sustainable Asia and the Pacific.* Manila

Over the next 5 years, ADB will continue to strengthen resilience and livability in Pacific DMCs through (i) **improved urban services**, (ii) **spatial planning**, and (iii) **climate change focus and environmental protection** with synergies across these three pillars (Table 1). Interventions that contribute to economic development and job creation shall be prioritized, given the importance of the economy for building and sustaining resilience. The crosscutting theme of climate change and the environment is emphasized in the road map because of its importance for the Pacific, and the urgency to take actions. Gender will continue to be mainstreamed, and there will be an increased focus on developing more proactive or transformative gender elements.

Table 1: Focus Pillars for Achieving the Vision

Dimension	Improved Urban Services	Spatial Planning	Climate Change and Environment (Crosscutting Themes)
Technical	• Long-term, integrated and gender-responsive, and capital expenditure and operating expenditure planning • Climate change adaptation • Adoption of technology for design, monitoring, and operations • Access to urban services (alignment with SDG 6)	• Multihazard disaster risk assessments • Risk-resilient and gender-responsive urban planning • Planning for urban services in informal areas • Formalization of informal areas • Improved information/data	• Citywide wastewater management solutions • Climate and disaster proofing and adaptation pathways • Climate change mitigation • Nature-based solutions • Reduced energy and greenhouse gas footprints • Integrated SWM • Coastal/ecosystem rehabilitation • Earth/soil conservation • Blue ocean initiatives
Enabling environment	• SOE strengthening • Capacity supplementation for utility management • Performance-based service contracts • Capacity building for disaster risk resilience in urban areas • Technology for improved customer service/accountability	• Capacity building on urban planning/data management/infrastructure planning • Improving interagency coordination	• Business plans for local and regional SWM • Institutional framework for regional SWM • Integrated water resource management planning • Capacity building on ocean health in coastal cities
Gender and inclusiveness	• Vulnerability mapping • Improving services for peri-urban or informal settlements • Effective partnerships with CSOs • Women's professional development • Behavior change campaigns	• Community engagement through a participatory approach • Addressing housing affordability, safety and security, and services for the poor	• Awareness campaigns for solid and liquid waste management • Participation of CSOs in solid waste collection
Financial	• Planning and advocacy for financial sustainability • Partnerships and mobilization of concessional financing	• Identifying resources/partnerships for implementation of urban development plans • Linking GIS with own-source revenue systems (e.g., property taxation)	• Identify opportunities for domestic and international private sector participation in SWM • Pursue climate change financing • Ecosystem services valuation

CSO = civil society organization, GIS = geographic information system, SDG = Sustainable Development Goal, SOE = state-owned enterprise, SWM = solid waste management.

Notes: The vision of the Pacific Urban Development, Water Supply, and Sanitation Division is to ensure that each intervention builds the resilience of people, institutions, the environment, and infrastructure through investments, knowledge, and partnerships.

Source: Asian Development Bank.

Sewerage system construction in Koror, Palau.

Pillar 1: Improved urban services. The focus subsectors under pillar 1 are:[13]

- urban water supply (core)
- wastewater management (core)
- urban planning (emerging)
- integrated solid waste management (emerging)
- urban flood management (emerging)
- urban disaster risk resilience (emerging)

- hygiene promotion (emerging)
- urban health (emerging)
- urban mobility (emerging)
- affordable housing (new)
- informal settlement upgrading (new)

Urban mobility includes transport planning, traffic management, street furniture, and improvement of nonmotorized circulation such as cycling and walking and gender considerations, and not only road resurfacing. Urban disaster risk resilience may include multipurpose shelter facilities for community spaces that can be used for emergencies.

---

[13] Designation of "core" means that the subsector is predominant in ADB's portfolio. "Emerging" means that ADB has supported a few interventions. "New" means that ADB has not yet supported such interventions in the last 5 years.

(i)   **Technical.** ADB shall continue ensuring that its investments are demand-driven, forward-looking, robust, resilient, and make use of appropriate technologies. ADB shall also continue to support its clients with the strategic phasing of investments based on long-term and integrated planning. A thorough life cycle approach to infrastructure planning is particularly vital, given the logistics challenges and high cost of materials and replacement parts, and the environmental conditions that lead to the rapid corrosion of some materials. Appropriate technologies will be introduced to support remote monitoring, decision-making, and increased operational efficiencies. Infrastructure plans will be aligned with risk-resilient spatial plans as much as possible to ensure that land vulnerabilities, growth patterns, and equity dimensions are addressed. Climate change adaptation measures will be explored for all investments and upstream hazard investigations carried out to ensure robust and long-lasting solutions.

> ### Box 2: Advocacy as Part of the COVID-19 Response
>
> The global coronavirus disease (COVID-19) pandemic is an opportunity to advocate on the importance of water, sanitation, and hygiene (WASH) to respond and build resilience to such shocks, and improve health outcomes of water supply and sanitation investments. ADB will increase its efforts to advocate for higher prioritization of resources for WASH in the Pacific and to strengthen linkages between WASH and public health. The technical assistance for Strengthening Water, Sanitation, and Hygiene Practices and Hygiene Behavioral Change in the Pacific will be implemented across the region until the end of 2023.[a]
>
> ---
>
> [a] Asian Development Bank. 2020. *Technical Assistance for Strengthening Water, Sanitation, and Hygiene Practices and Hygiene Behavioral Change in the Pacific*. Manila (TA 6551-REG).
> Source: Asian Development Bank.

(ii)   **Enabling environment.** All projects shall have a well-defined institutional strengthening component and seek to build, transfer, or provide adequate skills through either management contracts, funding of individuals to supplement capacity, secondment arrangements for utility staff to O&M contractors, or through performance-based service contracts. Appropriate technologies will be introduced to improve stakeholder engagement and customer service and to increase accountability to consumers.

(iii)   **Gender and inclusiveness.** The following shall be pursued to ensure that everyone, including women and those with increased vulnerabilities, benefit from the investments: (a) mapping and surveys to identify most vulnerable persons and develop targeted solutions; (b) sustainable expansion or improvement of services to peri-urban areas and informal settlements; (c) establishment of strong partnerships with civil society for project design and implementation, and mainstreamed into post-construction phase; (d) opportunities for professional development of women in the project agencies, including in management and STEM[14] fields; and (e) behavior change campaigns on hygiene, waste management, and water use to ensure maximum impact of investments. ADB will also increase efforts to (a) prevent and respond to gender-based violence, including through gender-responsive urban planning and designs; and (b) improve menstrual health and hygiene.

(iv)   **Financial sustainability.** ADB will work with its counterparts to develop and advocate for sustainable financing plans to ensure adequate resources for O&M from tariffs, taxes, and transfers. ADB shall support service providers to achieve greater operational efficiencies to reduce costs; while promoting tariffs that are affordable and increasingly recover costs. ADB shall continue to work with its clients to secure climate financing.

---

[14]   Science, technology, engineering, and mathematics.

Aerial shot of the commercial center in Koror, Palau.

Pillar 2: Improved spatial planning. Support for spatial planning is intended to help (i) identify suitable locations for transport and energy infrastructure as well as urban infrastructure; (ii) determine appropriate land uses based on projected climate change impacts, and physical, social, and economic considerations; and (iii) ensure affordable and safe housing and basic services for all, including vulnerable populations. ADB support will help (i) align economic, spatial, and infrastructure plans; (ii) build capacity of executing and implementing agencies for preparing, implementing, and updating spatial plans; (iii) establish mechanisms for effective interagency coordination and community engagement; and (iv) provide strategic inputs for transport and energy sectors. With mounting constraints for developable land, ADB will support select Pacific DMCs in a multihazard disaster risk assessment to inform the urban planning process and infrastructure planning, while supporting atoll countries with critical land-use decisions and considerations such as elevating or reclaiming land.[15] ADB will also explore opportunities to improve data collection system and geographic information system for planning, urban governance, and operations. This builds on ADB's support for spatial planning in Solomon Islands and Vanuatu.

---

[15] Currently, a multihazard disaster risk assessment is underway in Tonga to support the development of the Climate- and Disaster-Resilient Urban Development Strategy and Investment Plan as part of ADB's Integrated Urban Resilience Sector Project for Tonga. The assessment will provide an understanding of the hazards (weather and geophysical; extreme events as well as slow onset), and the exposure and the vulnerability.

Pillar 3: Climate change and environment. Central to the vision is building resilience to climate change and improving and protecting ecosystems, given their fragility, high impact on people's day-to-day lives, and strong linkages with tourism. Climate change adaptation pathways shall be supported and mitigation solutions shall be maximized where viable with the introduction of renewable energies. ADB shall support its clients to plan for more holistic solutions to solid waste challenges addressing the wide range of waste types and create opportunities for domestic private sector. Support for SWM at local, national, and regional scales also contributes to ADB's Action Plan for Healthy Oceans. ADB will continue the focus on waste management to minimize pollution, while ensuring that facilities are resilient to seismic and climate events. Citywide approaches to improving access to basic sanitation shall be pursued to help ensure that human waste is safely managed along the whole sanitation service chain; and potential links between sanitation and food security explored. Solutions that are complex and expensive to operate shall be minimized as multiple experiences have demonstrated that sustainability is a major challenge for sanitation; and efforts shall be made to strengthen institutional frameworks, O&M systems, and regulations for sanitation.[16] Catchment-wide and citywide stormwater management and urban flood mitigation projects will continue to be supported, while considering the viability of more nature-based solutions.

Project design. ADB will build on lessons from previous projects to help ensure smooth implementation and tangible and lasting results. Project designs should strive for innovation, integration, and complete solutions; but they should be based on a deep understanding of the government's capacity for interagency coordination, reforms, stakeholder engagement, implementation issues, and O&M.

How we work is as important as **what we do**. ADB's projects will strive for a service-oriented approach to the investments, leverage the private sector, and ensure high quality at entry, securing commitment for reforms early and adopting relevant technologies. These approaches shall help to create a conducive enabling environment for project implementation while maximizing the development impact. How ADB will work includes:

> **Box 3: Leveraging Power Sector Reforms for Advancing the Water Supply and Sanitation Sector**
>
> Through the energy sector, the Asian Development Bank (ADB) has supported the reform of several state-owned enterprises that are responsible for power, water supply, and/or sewerage. The Pacific Urban Development, Water Supply, and Sanitation Division will seek to leverage these gains for the water supply and sanitation (WSS) sector. ADB's investments in power supply have also contributed significantly to the performance of WSS systems as many depend on pumping. The experience of these state-owned enterprises with ADB project requirements should contribute to efficient and effective project implementation; and the strengthened capacity and existing commitment for reforms provide a conducive enabling environment to build on for the WSS sector. Such opportunities are present in the Federated States of Micronesia (Chuuk, Kosrae, and Pohnpei); Kiribati; the Marshall Islands (Ebeye); Nauru; Palau; and Tuvalu.
>
> Source: Asian Development Bank.

(i)   **Long-term partnerships.** A programmatic approach aligned with longer-term country sector plans allowing deeper engagement on policy issues will be pursued. Long-term partnerships with incremental gains should lead to more transformative results over time. ADB will work to strengthen relationships and develop multiyear pipelines in core countries, while seeking new opportunities in countries such as Nauru and Samoa. ADB will scale up its efforts to (a) advocate for urban development to be a national priority; (b) leverage gains achieved through power sector for state-owned enterprise reforms (e.g., for state-owned enterprises also responsible for water supply and/or sanitation); (c) strengthen policy and

---

[16]   ADB. 2018. *Leading Factors of Success and Failure in Asian Development Bank Urban Sanitation Projects*. Manila.

Water supply tank farm in Majuro, Marshall Islands.

regulatory frameworks; (d) strengthen institutional capacity for project implementation; and (e) build or supplement capacity for planning and operations.

(ii) **Service-oriented approach.** ADB will place an emphasis on service-oriented solutions in line with Sustainable Development Goal 6. A service-oriented approach requires resource-efficient planning for the best outcomes for end-users. It requires building professional O&M and asset management systems, and strengthening demand management. To support improvements in service quality, ADB may consider financing (a) performance-based service contracts and (b) renewal and rehabilitation of assets rather than wholesale asset replacement.

(iii) **Private sector participation.** Opportunities for private sector participation will continue to be sought, particularly for industry know-how, skills transfer, effective O&M, and financing. ADB will promote the use of design–build–manage or design–build–operate contracts to accelerate skills transfer and to prove functionality of investments.

(iv) **High quality at entry.** Each project will include a thorough risk management assessment building on lessons from other projects in the country and the subregion to ensure a robust project design. Designs will be matched with government priorities to ensure high ownership. High project readiness through project readiness financing facility or other mechanism will be sought for designs, procurement, gender elements, safeguards, and establishment of the project management units to avoid major design changes during implementation and start-up delays.[17] Independent review of engineering designs will be pursued for quality assurance.

---

[17]   Other mechanisms for preparing detailed designs include the use of project loan/grant for preparation of additional financing components.

(v) **Secure commitment for reforms.** ADB will promote enhanced participatory approaches and community engagement to promote beneficiary "buy-in" and build stakeholder ownership of projects and reform measures, especially related to tariffs and institutional arrangements. Champions at national, subnational, agency, and community levels will be identified to help ensure broad stakeholder understanding and support for a common direction. Achievement of early milestones of a longer-term reform road map will be pursued during project preparation to confirm commitment. Resources shall be allocated for individual consultants to provide support on a long-term, intermittent basis for implementation of reforms, advocacy, and capacity building.

(vi) **Increase use of technology.** Appropriate technologies will play an increasing role to modernize operations and reduce operating costs, enhance quality of data and designs, and improve customer services. The reliance on desalination technologies for water supply is likely to increase in most Pacific DMCs. Application of technologies such as "digital twin" will be explored. Adequate capacity or appropriate contract modalities to accompany the introduction and use of new technologies will be ensured.

ADB has set for itself several targets in line with the vision to scale up ongoing efforts, while embarking in some new directions (Table 2). Efforts will be made to develop a pipeline of projects in Fiji, PNG, and Solomon Islands, which have access to concessional loans in addition to the Asian Development Fund grants.[18]

Table 2: Strategic Targets, 2020–2025

Advisory/pipeline	Complete a study on potential interventions in affordable housing in at least 1 Pacific DMC
Advisory/pipeline	Identify potential interventions in urban mobility in at least 1 Pacific DMC.
Advisory/project preparation	Complete multihazard disaster risk analysis in at least 2 Pacific DMCs (sector- and project-specific)[a]
Advocacy/knowledge sharing	Support annual event of PWWA and regular webinars on utility management
Capacity building	Initiate at least 3 twinning arrangements
Financing	Secure the ADB Board of Directors approval for at least: 1 project with the main objective to improve services in peri-urban or informal areas 100% of projects having at least EGM categorization 2 projects including performance-based services contracts 2 projects with investments following a CWIS approach
Project preparation	Achieve high degree of readiness for a regional SWM project
Project preparation	Include a hygiene component in all water supply and sanitation projects

ADB = Asian Development Bank; CWIS = citywide, inclusive sanitation; EGM = effective gender mainstreaming; DMC = developing member country; PWWA = Pacific Water and Wastewater Association; SWM = solid waste management.

[a] To be carried out in coordination with ADB's climate specialists.

Source: Asian Development Bank.

---

18   Also, Vanuatu is eligible to borrow from ADB, but current government policy does not permit it because of fiscal pressures.

Tables 3 and 4 provide a snapshot of ongoing investments and advisory work and emerging opportunities. ADB shall continue advisory work to provide knowledge solutions (Finance+) and to inform new investment opportunities. Emerging opportunities have been identified through consultations with key counterparts during country programming, inputs of ADB staff, and project completion reports.

Table 3: Current and Potential Investment Opportunities

| Country | Current Portfolio | | | | | | | | | Emerging Opportunities |
	Urban water	Urban Sanitation	Peri-urban WSS	Hygiene Promotion	SWM	Flood Management	Urban Mobility	Urban DRR	Affordable Housing	
Cook Islands		X		X						SWM
Fiji	X	X	X			X				Urban mobility, affordable housing
Kiribati	X			X						
Nauru				X						Risk-resilient spatial planning, WSS, SWM
Niue										
Palau	X	X		X						
Papua New Guinea	X	X	X	X						
Marshall Islands	X	X	X	X	X					
Micronesia, Federated States of	X	X		X						
Samoa										Urban water
Solomon Islands	X	X	X	X		X	X	X		SWM
Tonga	X	X		X	X	X	X	X		Climate change pathway
Tuvalu	X	X		X		X				Peri-urban WSS, SWM
Vanuatu	X	X	X	X		X	X			Service contract for water supply
Regional		X?		X						SWM, WASH, water quality monitoring

DRR = disaster risk resilience; SWM = solid waste management; WASH = water, sanitation, and hygiene; WSS = water supply and sanitation.

Source: Asian Development Bank. 2020. *Annual Pacific Urban Update*. Manila.

Table 4: Priorities for Advisory Work and Capacity Building

Issue	Where We Are Working	Emerging Opportunities
Ongoing		
Multihazard disaster risk assessment	Tonga	All Pacific DMCs
SWM planning and management (city level)	Marshall Islands	Solomon Islands
Town/spatial planning	Palau, Solomon Islands, Vanuatu	Fiji, Nauru
Twinning – local government[a]	Fiji	Vanuatu
Twinning – water utility	Fiji, Samoa, Tonga, Solomon Islands	FSM, Palau, PNG, Marshall Islands
Water utility diagnostic reviews	Cook Islands, Solomon Islands, Vanuatu	All Pacific DMCs
Urban WSS sector planning and management	FSM, Kiribati, Palau, PNG, Marshall Islands, Solomon Islands, Tuvalu, Vanuatu	Nauru
Urban WSS sector policy and regulation	Fiji, Tuvalu	
Urban development policies	Vanuatu	Fiji
WSS master planning and feasibility studies	Kiribati, PNG, Solomon Islands, Vanuatu, FSM, Marshall Islands	Nauru
Water quality monitoring and data analysis	Vanuatu, Solomon Islands	All Pacific DMCs
Potential new areas		
Affordable housing	Not applicable	Fiji
Menstrual health and hygiene		PNG
Gender-based violence		
Regional SWM		Fiji
Sustainable tourism		Fiji
Upgrading of informal settlements		Fiji, Vanuatu
Urban mobility		Fiji

DMC = developing member country, FSM = Federated States of Micronesia, PNG = Papua New Guinea, SWM = solid waste management, WSS = water supply and sanitation.

[a] Could focus on SWM, financial management, and other aspects of local government operations.

Source: Asian Development Bank.

4 ONE ADB APPROACH

Children collecting water in Ebeye, Marshall Islands.

ADB will pursue greater collaboration and integrated solutions to complex challenges to achieve the objectives of ADB's Strategy 2030, the Livable Cities Operational Plan (OP4); the Pacific Approach; and ADB's 5-year targets for the urban development, water supply and sanitation sector in the Pacific (see Table 2). Collaboration will be sought with:

(i)   ADB Pacific Department's Energy; Social Sectors and Public Sector Management; and Transport and Communications divisions on areas such as energy efficiency, public health, municipal finance, and urban mobility;
(ii) other departments within ADB; and
(iii) with specific trust funds supported by ADB such as the Pacific Regional Infrastructure Facility and the Pacific Private Sector Development Initiative.

5 PROJECT FINANCE+ ACTIVITIES

Man fishing outside Suva, Fiji.

Finance+ activities and knowledge solutions are an integral part of this road map. Finance+ shall be pursued for areas such as climate change adaptation, financial sustainability, hygiene and sanitation promotion, nonrevenue water, performance-based service contracts, and other topics depending on clients' needs, demand, and ADB advantage. Gender mainstreaming will be a key consideration across all Finance+ activities.

Finance+ activities shall be achieved through:

(i)   analytical work linked to current and potential investments;
(ii)   mobilization of additional financial resources from other donors, development partners, and climate funds such as Australia Department of Foreign Affairs and Trade, European Union, Global Environment Facility, Green Climate Fund, New Zealand Ministry of Foreign Affairs and Trade and World Bank;
(iii) peer-to-peer exchanges (e.g., twinning on WSS, SWM, urban planning, municipal governance; knowledge webinars on utility management; and public health);
(iv)   establishment and deepening of relationships with knowledge partners such as renowned universities and think tanks; and
(v)   continued support for regional bodies such as the Pacific Water and Wastewater Association.

APPENDIX

DEMOGRAPHIC AND URBAN SERVICES DATA

Table A1.1: Data on Pacific Developing Member Countries
(by population)

Pacific Developing Member Country	Total Population (million)	Urban Population (million)	% Urban	Land Area (km²)	Number of Islands (approximate)
Papua New Guinea	7.934	1.040	13	462,840	600+
Fiji	0.903	0.492	55	18,333	300
Solomon Islands	0.606	0.141	23	5,750	6 major 900+ minor
Vanuatu	0.276	0.074	27	12,199	83
Samoa	0.196	0.037	19	2,842	9
Kiribati	0.116	0.049	42	726	33
Tonga	0.108	0.026	24	747	170
Federated States of Micronesia	0.106	0.024	23	701	607
Marshall Islands	0.053	0.039	74	181	29 atolls (each made up of many islets) and 5 islands
Palau	0.022	0.019	88	458	340
Cook Islands	0.021	0.016	75	240	15
Tuvalu	0.010	0.006	61	26	9 major islands 124 islands + islets
Nauru	0.010	0.010	100	21	1
Niue	0.002	0.0008	46	261	1
Total	10.36	1.97	19		

km² = square kilometer.

Source: Asian Development Bank.

Table A1.2: Drinking Water Estimates (1)

Country, Area, or Territory	ISO3	Year	Population ('000)	% Urban	National					Rural					Urban				
					At Least Basic	Limited (more than 30 mins)	Unimproved	Surface Water	Annual Rate of Change in Basic	At Least Basic	Limited (more than 30 mins)	Unimproved	Surface Water	Annual Rate of Change in Basic	At Least Basic	Limited (more than 30 mins)	Unimproved	Surface Water	Annual Rate of Change in Basic
Cook Islands	COK	2000	18	65	100	-	0	0	0.00	-	-	-	-	-	-	-	-	-	-
		2015	21	75	100	-	0	0		-	-	-	-		-	-	-	-	
Fiji	FJI	2000	811	48	95	-	3	2	-0.06	91	-	5	4	-0.12	99	-	1	0	-0.07
		2015	892	54	94	-	4	2		89	-	7	4		98	-	2	0	
Kiribati	KIR	2000	84	43	61	1	39	0	0.24	49	1	50	0	-0.29	77	0	23	0	0.84
		2015	112	44	64	1	35	0		44	1	55	0		90	0	10	0	
Marshall Islands	MHL	2000	52	68	-	-	-	-	-	-	-	-	-	-	-	-	-	-	-
		2015	53	73	78	21	1	0		99	0	1	0		70	28	2	0	
Micronesia (Federated States of)	FSM	2000	107	22	93	-	7	0	-0.28	92	-	8	0	-0.41	94	-	6	0	0.14
		2015	104	22	88	-	12	0		86	-	14	0		97	-	3	0	
Nauru	NRU	2000	10	100	95	0	5	0	0.33	-	-	-	-	-	95	0	5	0	0.33
		2015	10	100	100	0	0	0		-	-	-	-		100	0	0	0	
Niue	NIU	2000	2	33	99	-	1	0	-0.07	-	-	-	-	-	-	-	-	-	-
		2015	2	43	98	-	2	0		-	-	-	-		-	-	-	-	
Palau	PLW	2000	19	70	92	-	8	0	0.53	80	-	20	0	1.09	97	-	3	0	0.23
		2015	21	87	100	-	0	0		97	-	3	0		100	-	0	0	
Papua New Guinea	PNG	2000	5,371	13	37	2	20	42	-0.01	29	1	21	48	0.00	84	4	8	4	0.00
		2015	7,619	13	37	2	20	42		29	1	21	48		84	4	8	4	
Samoa	WSM	2000	175	22	93	2	5	0	0.17	94	2	3	0	0.03	89	0	10	0	0.69
		2015	193	19	96	2	3	0		95	2	3	0		99	1	0	0	
Solomon Islands	SLB	2000	412	16	80	6	10	4	-1.08	78	6	11	4	-1.46	90	3	5	1	0.00
		2015	584	22	64	4	17	15		56	5	20	19		90	3	5	1	
Tonga	TON	2000	98	23	98	-	2	0	0.09	99	-	1	0	0.07	97	-	3	0	0.16
		2015	106	24	100	-	0	0		100	-	0	0		100	-	0	0	
Tuvalu	TUV	2000	9	46	-	-	-	-	-	-	-	-	-	-	-	-	-	-	-
		2015	10	60	99	-	1	0		99	-	1	0		100	-	0	0	
Vanuatu	VUT	2000	185	22	82	1	11	7	0.59	78	1	13	8	0.64	96	0	4	0	0.25
		2015	265	26	91	1	2	6		87	1	3	8		99	0	0	0	
Average			167.5	44.8	85.9	3.3	8.0	4.6	0.04	78.6	1.9	13.0	7.2	-0.1	93.0	3.3	4.5	0.5	0.3

- = not available.

Note: The category of small island developing states (SIDS) used by the Joint Monitoring Programme is wider than the Asian Development Bank's Pacific developing member countries. The Joint Monitoring Programme SIDS category includes SIDS in Africa and Latin America and the Caribbean, as well as American Samoa, French Polynesia, Northern Mariana Islands, Singapore, Suriname, and Timor-Leste. The data for SIDS are indicative for the Pacific region.

Source: World Health Organization and United Nations Children's Fund (UNICEF) Joint Monitoring Programme for Water Supply, Sanitation and Hygiene. 2017. *A Snapshot of Drinking Water, Sanitation and Hygiene in the UNICEF East Asia & Pacific Region.* Geneva.

Table A1.3: Drinking Water Estimates (2)

Country, Area, or Territory	Year	National Proportion of Population Using Improved Water Supplies						Rural Proportion of Population Using Improved Water Supplies						Urban Proportion of Population Using Improved Water Supplies					
		Safely Managed	Accessible Premises	Available When Needed	Free from Contamination	Piped	Non-piped	Safely Managed	Accessible Premises	Available When Needed	Free From Contamination	Piped	Non-piped	Safely Managed	Accessible Premises	Available When Needed	Free From Contamination	Piped	Non-piped
Cook Islands	2000	-	85	100	-	69	31	-	-	-	-	-	-	-	-	-	-	-	-
	2015	-	87	100	-	75	25	-	-	-	-	-	-	-	-	-	-	-	-
Fiji	2000	-	78	95	-	-	-	-	60	-	-	-	-	-	97	-	-	-	-
	2015	-	69	94	-	87	7	-	37	-	-	74	15	-	96	-	-	97	1
Kiribati	2000	-	52	-	-	32	29	-	34	-	-	21	29	-	72	-	-	48	29
	2015	-	56	-	-	33	32	-	84	-	-	3	42	-	84	-	-	71	19
Marshall Islands	2000	-	-	-	-	-	-	-	-	-	-	0	-	-	-	-	-	-	-
	2015	-	74	-	-	11	88	-	98	-	-	0	99	-	65	-	-	15	83
Micronesia (Federated States of)	2000	-	66	-	-	-	-	-	66	-	-	-	-	-	69	-	-	-	-
	2015	-	33	-	-	-	-	-	61	-	-	-	-	-	71	-	-	-	-
Nauru	2000	-	95	-	-	-	-	-	-	-	-	-	-	-	95	-	-	-	-
	2015	-	99	-	-	68	32	-	-	-	-	-	-	-	99	-	-	68	32
Niue	2000	98	99	99	98	99	0	-	-	-	-	-	-	-	-	-	-	-	-
	2015	97	98	98	97	87	11	-	-	-	-	-	-	-	-	-	-	-	-
Palau	2000	-	84	-	-	92	0	-	74	-	-	80	0	-	88	97	-	97	0
	2015	-	95	-	-	100	0	-	93	-	-	97	0	-	95	100	-	100	0
Papua New Guinea	2000	-	19	-	-	20	18	-	16	-	-	13	18	-	43	80	-	71	17
	2015	-	19	-	-	20	18	-	16	-	-	13	18	-	43	80	-	71	17
Samoa	2000	-	92	95	-	86	9	-	93	-	-	85	11	-	88	-	-	90	0
	2015	-	94	97	-	82	15	-	93	-	-	84	13	-	98	-	-	77	23
Solomon Islands	2000	-	62	44	-	59	27	-	58	68	-	57	28	-	83	-	-	73	21
	2015	-	51	35	-	47	21	-	42	49	-	40	21	-	83	-	-	73	21
Tonga	2000	-	98	-	-	96	2	-	99	-	-	97	2	-	97	-	-	94	3
	2015	-	71	-	-	94	6	-	71	-	-	97	3	-	74	-	-	86	14
Tuvalu	2000	-	-	-	-	-	-	-	-	-	-	-	-	-	-	-	-	-	-
	2015	-	97	-	-	99	0	-	-	-	-	99	0	0	-	0	50	100	0
Vanuatu	2000	-	45	-	-	50	32	-	38	-	-	41	38	-	68	-	-	83	13
	2015	-	50	-	-	32	60	-	43	-	-	23	65	-	71	-	-	56	44
Average			71.8	85.7	97.5	65.4	21.0		61.9	58.5		51.3	23.6		80.0	71.4	50.0	76.1	18.7

- = not available.

Note: The category of small island developing states (SIDS) used by Joint Monitoring Programme is wider than the Asian Development Bank's Pacific developing member countries. The Joint Monitoring Programme SIDS category includes SIDS in Africa and Latin America and the Caribbean, as well as American Samoa, French Polynesia, Northern Mariana Islands, Singapore, Suriname, and Timor-Leste. The data for SIDS are indicative for the Pacific region.

Source: World Health Organization and United Nations Children's Fund (UNICEF) Joint Monitoring Programme for Water Supply, Sanitation and Hygiene. 2017. *A Snapshot of Drinking Water, Sanitation and Hygiene in the UNICEF East Asia & Pacific Region.* Geneva.

Table A1.4: Sanitation Estimates (1)

Country, Area, or Territory	ISO3	Year	Population ('000)	% Urban	National						Rural						Urban					
					At Least Basic	Limited (shared)	Unimproved	Open Defecation	Annual Rate of Change in Basic	Annual Rate of Change in Open Defecation	At Least Basic	Limited (shared)	Unimproved	Open Defecation	Annual Rate of Change in Basic	Annual Rate of Change in Open Defecation	At Least Basic	Limited (shared)	Unimproved	Open Defecation	Annual Rate of Change in Basic	Annual Rate of Change in Open Defecation
Cook Islands	COK	2000	18	65	92	0	7	1	0.37	-	-	-	-	-	-	-	-	-	-	-	-	-
		2015	21	25	98	0	2	-			-	-	-	-			-	-	-	-		
Fiji	FJI	2000	811	48	80	3	16	1	1.01	-0.03	71	3	25	1	1.63	-0.06	91	3	5	0	0.33	0.00
		2015	892	54	96	4	0	0			95	4	1	0			96	4	0	0		
Kiribati	KIR	2000	84	43	30	7	14	49	0.63	-0.94	21	3	22	55	0.76	-0.31	43	12	4	41	0.43	-1.71
		2015	112	44	40	8	17	35			32	4	14	50			49	14	22	15		
Marshall Islands	MHL	2000	52	68	-	-	-	-	-	-	-	-	-	-	-	-	-	-	-	-	-	-
		2015	53	73	87	0	2	11			66	0	4	30			95	0	2	4		
Micronesia (Federated States of)	FSM	2000	107	22	-	-	-	-	-	-	-	-	-	-			-	-	-	-	-	-
		2015	104	22	-	-	-	-			-	-	-	-			-	-	-	-		
Nauru	NRU	2000	10	100	66	31	3	1	-0.01	0.11	-	-	-	-	-	-	66	31	3	1	-0.01	0.11
		2015	10	100	66	31	1	3			-	-	-	-			66	31	1	3		
Niue	NIU	2000	2	33	100	0	0	0	-0.22	0.00	-	-	-	-			-	-	-	-		
		2015	2	43	97	0	3	0			-	-	-	-			-	-	-	-		
Palau	PLW	2000	19	70	85	0	15	0	1.02	0.00	67	0	33	0	2.20	0.00	92	0	8	0	0.52	0.00
		2015	21	87	100	0	0	0			100	0	0	0			100	0	0	0		
Papua New Guinea	PNG	2000	5 374	13	19	3	65	13	-0.01	0.00	13	3	70	14	0.00	0.00	55	9	32	4	0.00	0.00
		2015	7 619	13	19	3	65	13			13	3	70	14			55	9	32	4		
Samoa	WSM	2000	175	22	99	0	1	0	-0.14	0.00	98	0	1	0	-0.14	0.00	99	0	1	0	-0.09	0.02
		2015	193	19	97	0	3	0			96	0	4	0			98	0	2	0		
Solomon Islands	SLB	2000	412	16	21	3	13	63	0.69	-1.47	13	2	12	73	0.36	-1.53	62	12	17	9	0.91	0.02
		2015	584	22	31	5	23	41			18	2	29	50			76	15	0	9		
Tonga	TON	2000	98	23	89	1	10	0	0.32	0.00	86	1	13	0	0.45	0.00	99	1	0	0	-0.15	0.00
		2015	106	24	93	1	6	0			92	1	6	0			97	1	3	0		
Tuvalu	TUV	2000	9	46	-	-	-	-	-	-	-	-	-	-			-	-	-	-		
		2015	10	60	91	0	1	7			91	0	0	9			92	0	2	6		
Vanuatu	VUT	2000	185	22	53	17	28	2	0.03	-0.01	51	13	34	2	0.00	-0.03	61	32	7	0	0.00	0.07
		2015	265	26	53	18	27	2			51	13	34	2			61	32	6	1		
Average				43.0	70.9	5.6	13.4	10.5	0.3	-0.2	59.7	2.9	20.7	16.7	0.7	-0.2	77.7	10.3	7.4	4.9		

- = not available.

Note: The category of small island developing states (SIDS) used by the Joint Monitoring Programme is wider than the Asian Development Bank's Pacific developing member countries. The Joint Monitoring Programme SIDS category includes SIDS in Africa and Latin America and the Caribbean, as well as American Samoa, French Polynesia, Northern Mariana Islands, Singapore, Suriname, and Timor-Leste. The data for SIDS are indicative for the Pacific region.

Source: World Health Organization and United Nations Children's Fund (UNICEF) Joint Monitoring Programme for Water Supply, Sanitation and Hygiene. 2017. *A Snapshot of Drinking Water, Sanitation and Hygiene in the UNICEF East Asia & Pacific Region.* Geneva.

Table A1.5: Sanitation Estimates (2)

Country, Area, or Territory	Year	National — Proportion of Population Using Improved Sanitation Facilities (excluding shared)							Rural — Proportion of Population Using Improved Sanitation Facilities (excluding shared)							Urban — Proportion of Population Using Improved Sanitation Facilities (excluding shared)						
		Safely Managed	Disposed in Situ	Emptied and Treated	Wastewater Treated	Latrines and Other	Septic Tanks	Sewer Connections	Safely Managed	Disposed in Situ	Emptied and Treated	Wastewater Treated	Latrines and Other	Septic Tanks	Sewer Connections	Safely Managed	Disposed in Situ	Emptied and Treated	Wastewater Treated	Latrines and Other	Septic Tanks	Sewer Connections
Cook Islands	2000	-	-	-	-	-	-	-	-	-	-	-	-	-	-	-	-	-	-	-	-	-
	2015	-	-	-	-	-	-	-	-	-	-	-	-	-	-	-	-	-	-	-	-	-
Fiji	2000	-	-	-	-	-	-	-	-	-	-	-	-	-	-	-	-	-	-	-	-	-
	2015	-	-	-	-	-	-	-	-	-	-	-	-	-	-	-	-	-	-	-	-	-
Kiribati	2000	-	-	-	-	1	20	10	-	-	-	-	0	16	5	-	-	-	-	2	25	16
	2015	-	-	-	-	8	22	10	-	-	-	-	7	20	5	-	-	-	-	8	25	16
Marshall Islands	2000	-	-	-	-	-	-	-	-	-	-	-	-	-	-	-	-	-	-	-	-	-
	2015	-	-	-	-	15	28	44	-	-	-	-	29	36	0	-	-	-	-	9	25	60
Micronesia (Federated States of)	2000	-	-	-	-	-	-	-	-	-	-	-	-	-	-	-	-	-	-	-	-	-
	2015	-	-	-	-	-	-	-	-	-	-	-	-	-	-	-	-	-	-	-	-	-
Nauru	2000	-	-	-	-	30	20	16	-	-	-	-	-	-	-	-	-	-	-	30	20	16
	2015	-	-	-	-	30	20	16	-	-	-	-	-	-	-	-	-	-	-	30	20	16
Niue	2000	-	-	-	-	15	78	7	-	-	-	-	-	-	-	-	-	-	-	-	-	-
	2015	-	-	-	-	12	78	7	-	-	-	-	-	-	-	-	-	-	-	-	-	-
Palau	2000	17	17	0	0	0	34	51	-	-	-	0	0	55	12	12	12	0	0	0	24	68
	2015	20	20	0	0	0	39	61	-	-	-	0	0	88	12	16	16	0	0	0	32	68
Papua New Guinea	2000	-	-	-	2	12	4	3	-	-	-	0	11	1	1	-	-	-	11	14	21	20
	2015	-	-	-	2	12	4	3	-	-	-	0	11	1	1	-	-	-	11	14	21	20
Samoa	2000	-	-	-	-	15	83	0	-	-	-	-	17	82	0	-	-	-	-	9	90	0
	2015	-	-	-	-	10	87	0	-	-	-	-	10	86	0	-	-	-	-	7	91	0
Solomon Islands	2000	-	-	-	-	10	11	-	-	-	-	-	9	4	-	-	-	-	-	15	47	-
	2015	-	-	-	-	19	13	-	-	-	-	-	16	3	-	-	-	-	-	29	47	-
Tonga	2000	-	-	-	-	9	80	0	-	-	-	-	9	77	0	-	-	-	-	9	90	0
	2015	-	-	-	-	14	80	0	-	-	-	-	16	77	0	-	-	-	-	7	90	0
Tuvalu	2000	-	-	-	-	-	-	-	-	-	-	-	-	-	-	-	-	-	-	-	-	-
	2015	9	9	0	0	10	8	73	14	14	0	0	13	15	63	6	6	0	0	7	4	81
Vanuatu	2000	-	-	-	-	32	14	7	-	-	-	-	41	3	7	-	-	-	-	6	46	9
	2015	-	-	-	-	30	16	8	-	-	-	-	41	3	7	-	-	-	-	6	46	9
Average			15.33	0	0.8	14.2	36.95	17.6		14	0	0	14.38	35.438	8.07		11.33	0	4.4	11.22	42.444	24.9

- = not available.

Note: The category of small island developing states (SIDS) used by the Joint Monitoring Programme is wider than the Asian Development Bank's Pacific developing member countries. The Joint Monitoring Programme SIDS category includes SIDS in Africa and Latin America and the Caribbean, as well as American Samoa, French Polynesia, Northern Mariana Islands, Singapore, Suriname, and Timor-Leste. The data for SIDS are indicative for the Pacific region.

Source: World Health Organization and United Nations Children's Fund (UNICEF) Joint Monitoring Programme for Water Supply, Sanitation and Hygiene. 2017. *A Snapshot of Drinking Water, Sanitation and Hygiene in the UNICEF East Asia & Pacific Region*. Geneva.

www.ingramcontent.com/pod-product-compliance
Lightning Source LLC
Chambersburg PA
CBHW050058220326
41599CB00045B/7457